BASICS TO BUSINESS

Minding Your Business With EXCEL

BASICS TO BUSINESS

Minding Your Business With EXCEL

Printed in the United States of America

ISBN: 1-4107-8537-8 (Paperback)

ISBN : 1-4107-8536-X (e-book)

This publication is designed to provide accurate and authoritative information about the subject matter covered. It is sold with the understanding that neither author nor publishers are engaged in rendering legal, accounting, or other professional service. If legal advice or other expert assistance is required, the services of a competent professional should be sought.

1stBooks-rev 11/17/03

About the Author

The author, S. D. Moore was born in Michigan. She later entered the United States Air Force and worked in the field of Human Services.

After an early retirement she kept quite busy. She hosted a self-improvement radio show, taught numerous classes in Excel, Word, PowerPoint, and Leadership. She worked for the U.S. Army as a Strategic Planner, completed a dual master's degree in Management and Human Resources Development, and patented an invention. She has earned several honors to include the "Who's Who Among Colleges and Universities" award in 2001.

To become an expert in her field, S. D. Moore elected to pursue a Doctorate's degree in Education with a major in Organizational Leadership. Her personal motto:

"Keep it Simple"

CONTENTS

In Memory of My Beloved Mother Shirley,
May You be Blessed with True Happiness.

INTRODUCTION

Welcome to a simple and inexpensive way of learning to use Excel in your small business. Basics to Business, guides you through quick and easy lessons to help you use and apply the various functions of Excel in your small business.

This step by step guide is divided into two parts: Part I The Basics and Part II Business Applications. You'll need a computer with the Excel application installed on it, an elementary understanding of arithmetic, and a "can do" attitude.

Now let's start with the basics.

Part I The Basics

Open the Excel application and study each menu command in the window. Starting with FILE, click on each of the command buttons at the top of the window (screen). Note what each command does. Roll the mouse over each icon to reveal what action each icon performs.

Now make sure that the frequently used shortcut buttons are displayed. Display the shortcut buttons by clicking on the VIEW command. A submenu will pop-up; roll your mouse down the list and select TOOLBARS. A new list will display. Select STANDARD, FORMATTING, and DRAWING (if there is a check to the left of a command, the item is already selected). Your screen should look like the following:

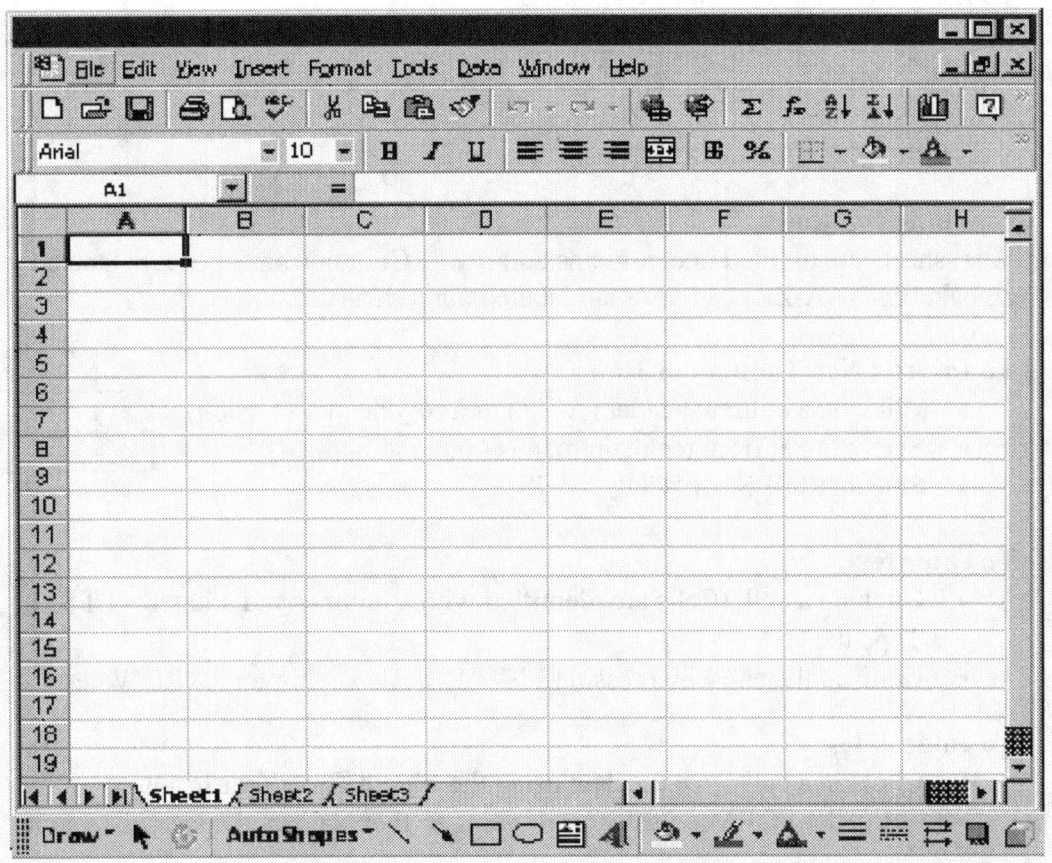

Before we go any further, DON'T FORGET TO SAVE YOUR DATA OFTEN.

TO SAVE A DOCUMENT: Click the File Command, Select SAVE AS, type the name of your document in the space titled FILE NAME, then click the save button.

* DO NOT USE THE SHORTCUT ICON UNTIL YOU'VE FOLLOWED THE PREVIOUS STEPS.

CREATING WORKSHEETS

What are workbooks and worksheets?

Workbooks In Excel, a workbook is a file that contains pages of data. Each workbook can contain many worksheets. You can organize various kinds of related information in a single file.

Worksheets Use worksheets to list and analyze data. If you need additional worksheets, Click the INSERT Command button from the main menu; select Worksheet. Worksheets will insert one tab to the left at the bottom of the screen.

Sheet tabs The names of the sheets appear on tabs at the bottom of the workbook window. To move from sheet to sheet, click the sheet tabs.

SHEET 1	SHEET 2

Columns and Rows
Worksheets are divided into columns and rows. Columns have headers labeled alphabetically (A-ZZ) and rows are labeled numerically.

To Insert a New Column or Row
- Click on a column header (A-ZZ) or a row header (1-1000)
- Select Insert from the main menu command bar.
- Select "ROWS" or "COLUMNS"

To enter text:
- Click on a cell, (Cells are identified with a number and a letter A1/F8)
- Type the text.
- Confirm the entry by pressing ENTER, TAB key, or an ARROW key.

To enter dates:
- Click on a cell Type the date using the Month/Date/Year format
- OR under FORMAT select cell, number, then date, and appropriate style.

EDITING WORKSHEETS

To edit data within a cell:
- Click on the cell that has the data to be edited.
- Either Bold, Italicize, Underline, Align, Delete
- OR Simply type the preferred data into the cell
- OR If you want to change part of the data in the cell, double click on the cell, select or highlight the specific data then type over it.

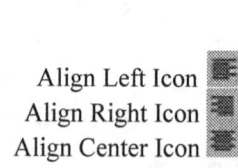

Align Left Icon
Align Right Icon
Align Center Icon

Bank in United	Bank of United

Cut, Copy, or Paste data into a cell:
- Click on the cell with the information to be edited.
- **Cut or Copy** the information by clicking on the cut or copy icon (this information will be placed on the clipboard).
- **Paste** the information from the clipboard by selecting the desired cell that you would like to place the information. Click on the cell then click the paste icon.

To copy data using drag and drop:
- Select the cell(s) you want to copy
- Position the mouse pointer on any border of the selected cell(s). Press and hold the Control key and the left button of the mouse and move with a sweeping motion to the destination cell(s) above, below, diagonal, or beside.
- Release the mouse button OR

To copy the same data several times:
- Select the cell(s) you want to copy
- Position the mouse pointer on the handle at the bottom right of the cell. Press and hold the left button of the mouse and move with a sweeping motion to the destination cells above, below, or beside.
- Release the mouse button.

To Insert, Delete or Clear data, cells, rows, or columns:
- Click on the tab at the top of the column (A) or at the end of the row.
- Click on the word **EDIT** on the command menu bar.
- Select **DELETE**, the data and format settings will be deleted from the area selected.
- The **CLEAR** command will allow greater flexibility with clearing specific items.
- **To reverse or redo actions:** Select undo or redo from the EDIT menu.

FORMATTING WORKSHEETS

To change margins:
- Click on File, Page Setup
- Click on the margins tab, then click on the appropriate arrows or type the desired margins in the text boxes for the top, bottom, left, and right margins
- Click on the OK button.

To center data on the page:
- Click on File, Page Setup
- Click on the Margin tab, next turn on the checkboxes for horizontally and/or vertically, as necessary Click on the OK button.

To position the page as a portrait ☐ **or landscape** ☐ **select page setup, then page.**

TO CREATE CUSTOM HEADERS AND FOOTERS:
- Click on File, Page Setup
- Click on the Header/Footer tab, then click on Custom Header or Custom Footer
- Type desired text in either the left, center, or right pane, then click the OK button
- To change the style of the text
- Highlight the text then select font, style, and size
- Click on the OK button.

Merge and center text:
- Click on cell containing text that you wish edited.
- Click on the Merge-and-Center ICON

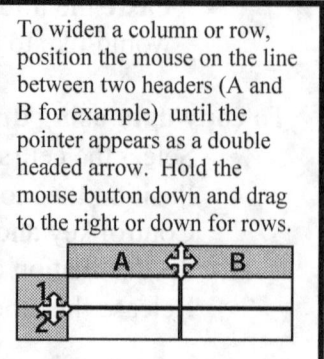

To widen a column or row, position the mouse on the line between two headers (A and B for example) until the pointer appears as a double headed arrow. Hold the mouse button down and drag to the right or down for rows.

Formatting the Cell width and height:
- If the column isn't wide enough your data will appear ######
- Click on the word **FORMAT** on the command menu bar.
- Select (Column) >then select the word (width).
- Type in the number designating the desired width of the cell.
- To increase the width of an entire row, select the specific cells to be resized then click on the word Format on the command menu bar, Select column, then Width, and type in the number designating the desired width of the row.
- **Width and Height** can also be changed by positioning the pointer on the line between the two letters at the top of the column or between the two numbers at the beginning of the row. (Example ⊕ ⇦⇨ between the A and B columns or between rows 4 and 5).

Sorting
Sorting allows you to alphabetize or categorize a list of data, contacts, or customers. To sort a list simply select all cells, however, DO NOT INCLUDE the headings at the top of each column. Select DATA on the main menu bar, then select SORT, check ASCENDING to alphabetize from A-Z, Check NO HEADER ROW, then click OK.

Filtering
Filtering allows you to locate and isolate certain data within a worksheet. For example if you want update information in your contact list on all companies from the state of New Jersey you would apply a filter to single-out companies from that state only. You would use this to locate and isolate other items such as names or food items, etc.
- Select cell the first cell, Click on DATA, Select Auto Filter, Click the down arrow, select item.
-To remove filter click on DATA, then select Auto Filter.

Freezing Panes
Freeze Panes is a great tool to use if you are creating a long list and you need to continuously view the headings at the top of each column as you input your data.
-Select the first cell just below the row that you want to be displayed (A8 for example), Click on WINDOW on the main menu command bar, and then select Freeze Panes.

Complete the following exercises: EX#1: Enter the following text or your own data; SORT in Ascending Order: Remember to select all of the cells BELOW THE HEADINGS at the top of each column.

EXERCISE 1 (Alphabetize/Sort) Save as: EXCEL LIST

792 Columbus	S. King Sister Henny Penny		Brother Kelly
M. King	Bond	Mr. Lanny King	S. King
Munster	Penny	D. Summers	1600 Preston

EXERCISE 2 (Alphabetize/Sort and Filter the data below)
Review how to sort and filter data on page 7.

1. In cell **C1** merge and center the following: THE DONNER COMPANY'S CONTACTS
2. In cell **A2** type LAST
3. In cell **B2** type FIRST
4. In cell **C2** type ADDRESS
5. In cell **D2** type PHONE
6. In cell **E2** type FAX
7. Enter the following data under the appropriate heading:

Herman D. Monster, 1313 Mockingbird CT, Nowhere, NJ, 61616 (212), 555-1313

Elvis King, 202 Star Dr., Love Me Tender, TN 31155, (333) 411-8989, FAX (333) 411-8900

Mark L. King, 8145 Dream Drive, Freedom, GA, (645) 311-7777, FAX (645) 311-7785

Jack Bond, 006 Spy Street, Rubies Are Forever, NV (702) 535-0007, FAX (702) 535-9007

Ms. Cash Penny, 12 Espionage Ave., Dr. No, NV (702) 535-7007, FAX (702) 535-8007

Henny Penny, 88 Hunk of Bread Ave, Loafin, LA, (622) 333-2332, FAX (622) 333-2300

Lanny King, 20 Glass Ct, Newsy, NJ, (411) 757-4111, FAX (411) 757-4115

Buffy Summers, 701 Slayer Street, Sunny Dale, CA 99991 (202) 616-5601

Steve King, 606, Nightmare LN, Misery, ME, (999) 919-9898 FAX (999) 919-9800

Bobby Kelly, 222 I Wish Ave., Downlow, D.C., (222) 777-4477 FAX (222) 777-4407

Roger Ness, 844 Prince Place, Artists, AZ, (404) 632-1401 FAX (404) 632-1425

Gladwin Night, 1979 Leaving St, Midnight Train To, GA (414) 272-2525 FAX (414) 272-2500

SAVE AS: CUSTOMER CONTACT LIST

WORKING WITH COLOR AND PICTURES

TO APPLY COLOR TO TEXT OR DATA:
- Click on the cell to be edited or double click and highlight the text or data
- Click on the DOWN ARROW of the FONT COLOR icon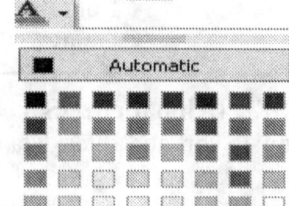
- Select the desired color from the palette
- Voila! The color of the data is changed.

TO INSERT A TEXTBOX
- Click on the TEXTBOX icon
- Draw a textbox
- Click on the frame of textbox; now you can change the color and or style of the contents in the textbox
- If you desire to change particular data inside the box, Double click inside the textbox, Select the data, and change its style and or color.

TO INSERT WORDART:
- Click on the WORDART icon A on the drawing toolbar at the bottom of the screen
- Select the style, click OK
- Type desired text, click OK
- Edit the color and shape on the Word Art editing menu.

TO APPLY COLOR TO A CELL, COLUMN, OR ROW:
- Click on the cell or highlight the column or row to be filled with color
- Click on the DOWN ARROW of the FILL COLOR icon (paint can)
- Select the desired color from the palette
- Voila! The color of the cell(s) is changed.

TO APPLY BORDERS AND ADD COLOR TO LINES:
- Select the cell or cells to be edited
- Click on the down arrow next to the borders icon ⊞ Select the preferable style from the menu
- OR Select the DRAW BORDERS option from the menu under the borders icon. Use the drawing tool to draw border lines in black or click the icon to draw lines in color.
- OR Click on the FORMAT command, select cell, click on the border tab; Select border, line style, and color from the menu; Click OK to complete the edit.

INSERT CLIPART
- Click on the Insert Clip Art icon
- OR click the INSERT command on the main menu
- Select picture, and then CLIP ART
- Search through and select a picture, click on the picture to insert it in the worksheet.
- Re-size by using the mouse to pull on one of the small boxes surrounding the picture.

PRINTING WORKSHEETS

How to print preview a worksheet

- Click on the Print Preview button in the Standard toolbar
- Click on the Close button.

To adjust page breaks:
- Click on View, Page Break Preview
- Drag the dashed blue lines to new locations
- Click on View, Normal.

To remove page breaks:
- Click on View, Page Break Preview
- Click on the page break you want to remove
- Click on Insert, Remove Page Break
- Click on View, Normal.

To change print size:
- Click on File, Page Setup
- Click on the Page tab, then click on the appropriate arrows in the Scaling section, or type the desired scale in the Adjust to text box
- Click on the OK button.

To print an entire worksheet:
- Click on the Print button on the Standard toolbar,
- OR Click on File, Print
- Select the appropriate printer, then type a number in the Number of copies text box
- Click on the OK button.

To print a selected area of a worksheet:
- Select the range of cells you want to print
- Click on File, Print Area, Set Print Area
- Click on the Print button.

To Print Column and Row Headings (column A – row 1):
- Click on the FILE command
- Select PAGE SETUP
- Click on the SHEET tab
- Check Row and Column Headings; then click OK

To clear a print area:
- Click on File Print Area, Clear Print Area.

Working with Basic Formulas

Creating formulas:

- Click on the cell in which you will enter the formula
- **Type an equal sign = then the word SUM**
- Inside of a parenthesis () Type the identifiers of selected cells (A1), an arithmetic operator. Example: =A1+A2 or =B3-B4-B5 or =sum(F1-C20) or =sum(E2+C9+D9+E9+F22) or
- By Column =average(F3:F18) or By Row =average(A12:F12)
- Continue as necessary until the formula is complete
- Hit the enter key to confirm the formula
- Or to **add** data in a column, click the auto-sum icon

Arithmetic Operators

+ (Plus sign)	Addition	3+3
− (Minus sign)	Subtraction	3-1
* (Asterisk)	Multiplication	3*3
/ (Forward slash)	Division	3/3
: (Range connector)	A1 through A10	A1:A10

To enter a function using a Formula Palette:
- Click on the cell in which you will enter the function
- Type an equal sign (=)
- Display the Function drop-down list, then click on the appropriate function name
- Enter the arguments by selecting the cells you want to include
- Click on the OK button.

To enter a function by typing:
- Click on the cell in which you will enter the function
- Type an equal sign (=)
- Type the function name and an opening parenthesis
- Type the range reference for the cells you want to include, using a colon to separate the first and last cell reference of the range.
- Type a closing parenthesis
- Press <Enter>.

EXERCISE 3: BASIC FORMULAS

Enter ALL OF THE DATA before calculating. Calculate the first 3 rows of each problem then DRAG-COPY the formula into the cells below.

To drag-copy: position the mouse's pointer (arrow) at the bottom right corner of the cell until the pointer changes from an ↗ to a + Hold the right button of the mouse down then drag down into or through the cell or cells below.

ADD THE FOLLOWING			TOTAL (Formula)
2	2		
15	15		
200	55		
88	44		
20	2046		

SUBTRACT THE FOLLOWING			TOTAL (Formula)
22	21		
164	25		
2500	55		
896	343		
888	1000		
1000	888		

MULTIPLY THE FOLLOWING			TOTAL (Formula)
29	30		
58	3		
11	12		
20	2046		
2500	55		
777	10		

DIVIDE THE FOLLOWING			TOTAL (Formula)
20	5		
800	20		
43	4		
533	27		
99	3		
2045	5		

AVERAGE THE FOLLOWING				TOTAL (Formula)
80	75	100		
65	72	75		
94	88	85		
100	98	90		
70	66	79		
90	90	82		

To apply the Currency format:
- Click on the Currency Style button in the Formatting toolbar, or $
- Right-click on the cell(s) you want to format, then click on Format Cells
- Click on Accounting in the Category list box
- Specify the number of decimals and the currency symbol if necessary
- Click on the OK button.

To apply the Percent format:
- Click on the Percent button in the Formatting toolbar %
- OR Type in a number with a percent symbol, or
- Right-click on the cell(s) you want to format, then click on Format Cells
- Click on Percentage in the Category list box
- Specify the number of decimals if necessary
- Click on the OK button

To apply the Number format:
- Click on the Increase Decimal button on the Formatting toolbar .00 +.0
- OR Right-click on the cell(s) you want to format, then click on Format Cells.

To apply the Comma format:
- Click on the Comma button in the Formatting toolbar. ,
- To apply the Date format:
- Right-click on the cell(s) you want to format, then click on Format Cells
- Click on Date in the Category list box
- Click on a format in the Type section
- Click the OK button.

NOTES:_____

_____

_____

_____

_____

Part II Business Applications

In this next section we will cover:

- Creating Charts
- Mixing Formulas
- Linking Formulas in Business Applications
- Creating Forms

Remember, the key to good management is preparation. Create a (formula) template in advance and plug-in the data later. This prevents managers from playing "catch-up" at the end of the month.

It is preferable to develop a record keeping system within the same workbook (document). **Remember, if you need additional worksheets, Click the INSERT Command button from the main menu; select Worksheet. Worksheets will insert one tab to the left at the bottom of the screen. To re-arrange the worksheets, click on the worksheet that you want to move; hold down the left button of the mouse and drag the worksheet to the preferred area in the sequence.**

Alright, we've covered the basics, let's get ready for Part II. Fill a glass with your favorite non-alcoholic beverage, fill a small bowl with a tasty snack, place on a serving tray, don't forget a napkin (don't want mess up the keyboard), position yourself in front of the computer; and let's get down to business.

*Please note that the exercises are simplistic models to better illustrate the use of Excel in basic business record keeping. Some businesses may require broader record keeping details.

MIXING FORMULAS

Mixing Formulas: PAYROLL, SALES REPORTS, AND BOOKKEEPING
In Basic Excel we used simple formulas to calculate basic equations. In this section we will mix formulas to setup and perform familiar office tasks.

WEEKLY PAYROLL
For this exercise, place the formula in cells D2, F2, and G2.
- **In cell D2 – Multiply cell B2 by cell C2**

CALCULATATE THE GROSS =sum(B2*C2)

	A	B	C	D	E	F	G
1	NAME	HOURLY	TOTAL HRS	GROSS	TAXES	WKLY TAXES	NET
2	Jim	$8.00	40	$320.00	11%	$35.20	$284.80
3	Mike	$16.00	35	$560.00	11%	$61.60	$498.40

CALCULATE THE TAXES =sum(D2*.11)

	D	E	F
1	GROSS	TAXES	WKLY TAXES
2	$320.00	11%	$35.20
3	$560.00	11%	$61.60

In cell F2 - Multiply cell D2 by .11
NOTE THAT PERCENTAGES MUST BE ENTERED INTO THE EQUATION AS A <u>NUMBER</u> NOT AS A CELL.

CALCULATE NET WEEKLY PAY =sum(D2-F2)

D	E	F	G
GROSS	TAXES	WKLY TAXES	NET
$320.00	11%	$35.20	$284.80
$560.00	11%	$61.60	$498.40

In cell G2 - Subtract cell F2 from cell D2
*Be careful not to transpose cell references

WEEKLY PAYROLL – Put it All Together

	A	B	C	D	E	F	G
1	NAME	HOURLY	TOTAL HRS	GROSS	TAXES	WKLY TAXES	NET
2	Jim	$8.00	40	$320.00	11%	$35.20	$284.80
3	Mike	$16.00	35	$560.00	11%	$61.60	$498.40

PAYROLL EXERCISE #4

Calculate the Bi-weekly salaries of 10 employees

BI WEEKLY PAYROLL						
	HOURLY	Total Hours	GROSS WEEKLY SALARY	TAX	TOTAL TAX	NET SALARY
MANAGER	20	80		18%		
SUPV	12	80		18%		
ADMIN	6	80		18%		
ADMIN	7	70		18%		
JANITOR	9	80		18%		
TRAINEE	11.5	40		18%		
TRAINER	15	80		18%		
TRAINER	14	70		18%		
TRAINER	14	80		18%		
TRAINER	14	40		18%		

There are 26 Bi-weekly pay periods in a year

ANNUAL PAYROLL						
	HOURLY	Projected Bi-weekly Hours	ANNUAL GROSS SALARY	TAX	TOTAL TAX	NET SALARY
MANAGER	20	80		18%		
SUPV	12	80		18%		
ADMIN	6	80		18%		
ADMIN	7	70		18%		
JANITOR	9	80		18%		
TRAINEE	11.5	40		18%		
TRAINER	15	80		18%		
TRAINER	14	70		18%		
TRAINER	14	80		18%		
TRAINER	14	40		18%		

NOTES:_____

_____

_____

_____

_____

PURCHASES

Sum the total amount purchased.
Subtract the total amount sold from the starting balance.
Merge and Center "December Purchases" also insert a header titled "Purchases"

DECEMBER PURCHASES					
			STARTING BALANCE	$	25,000.00
			ENDING BALANCE	$	**Formula**
DATE	**PRODUCT**	**AMOUNT**			
8-Dec-03	AC Compressor Belt	$ 500.00			
9-Dec-03	Holiday Decorations	$ 200.00			
9-Dec-03	AC Compressor	$ 2,300.00			
12-Dec-03	Basic Office Supplies	$ 65.00			
15-Dec-03	Desk	$ 900.00			
16-Dec-03	Laptop Computer	$ 1,200.00			
17-Dec-03	Projector	$ 1,600.00			
18-Dec-03	Projector Bulb	$ 200.00			
18-Dec-03	Basic Office Supplies	$ 55.00			
19-Dec-03	Training Tapes	$ 600.00			
20-Dec-03	Promotional Mugs	$ 1,250.00			
21-Dec-03	Basic Office Supplies	$ 49.50			
22-Dec-03	Company Sign	$ 2,200.00			
22-Dec-03	Desk	$ 425.00			
	TOTAL AMOUNT	**Formula**			

NOTES:_____

_____

_____

_____

_____

SALES

Sum the total amount sold.
Add the total amount sold to the starting balance.
Merge and Center "December Sales" also insert a header titled "Sales"

DECEMBER SALES				
			STARTING BALANCE	$ Insert ending balance from purchases
			ENDING BALANCE	$ Formula
DATE	**PRODUCT**	**AMOUNT**		
8-Dec-03	Seminar	$ 1,100.00		
9-Dec-03	Stress Mgt Class	$ 400.00		
9-Dec-03	4 Hour Seminar	$ 550.00		
12-Dec-03	Counseling	$ 75.00		
15-Dec-03	Business Consult	$ 250.00		
16-Dec-03	Seminar	$ 1,100.00		
17-Dec-03	Conflict Class	$ 600.00		
18-Dec-03	Seminar	$ 1,100.00		
18-Dec-03	Counseling	$ 75.00		
19-Dec-03	Leadership Seminar	$ 5,000.00		
20-Dec-03	Counseling	$ 75.00		
21-Dec-03	Dealing w/Difficult	$ 600.00		
22-Dec-03	Counseling	$ 75.00		
22-Dec-03	Business Consult	$ 175.00		
	TOTAL AMOUNT	**Formula**		

NOTES:_____

_____

_____

_____

_____

A SIMPLE PROJECTED ANNUAL BUDGET

Setting up the budget requires more steps, but it's not that difficult. Basically you allocate funds for weekly, monthly, and or annual expenses; i.e., payroll, utilities, rent, etc. For this exercise, add the sums of each category. The ending balance is the capital minus total costs.

Employee Expenses	Gross Annual	NOTES:
Employee's Title	Salary	
MANAGER		
SUPV		
ADMIN		
ADMIN		
JANITOR		
TRAINEE		
TRAINER		
TRAINER		
TRAINER		
TRAINER		
Total Employee Expenses	**SUM**	
Marketing Expenses	**Amount**	
TV ADS	$15,000	
NEWSPAPER AD	$8,000	
FLYERS	$3,000	
BROCHURES	$6,000	
Total Marketing Expenses	**SUM**	
General Expenses		
Rent	$45,000.00	
Utilities	$5,000.00	
Assessment Tools	$7,000.00	
Computer Equip Replacement	$5,000.00	
Office Supplies	$6,000.00	
Cleaning Supplies	$2,000.00	
Furniture	$2,000.00	
Total General Expenses	**SUM**	
		STARTING CAPITAL: **$180,000**
Total Costs	**FORMULA** *Add sums	EST. ENDING CAPITAL $FORMULA Starting Capital minus total costs

CREATING CHARTS

CHARTS

Charts allow the presenter to graphically explain data and figures. A viewer can get a better picture of the data being presented. If you look at the example below at first glance it's difficult to determine which department had the highest overall sales. However, if you look at the chart it's easier to make that determination.

To create a chart, simply type data into an EXCEL Worksheet. Remember that the **X** AXIS runs along the bottom of the chart and the **Y** AXIS runs along the right or left side of the chart. The **LEGEND is** usually square shaped and filled with a key to further explain the **X** AXIS (at the bottom of the chart). When you use the chart wizard, 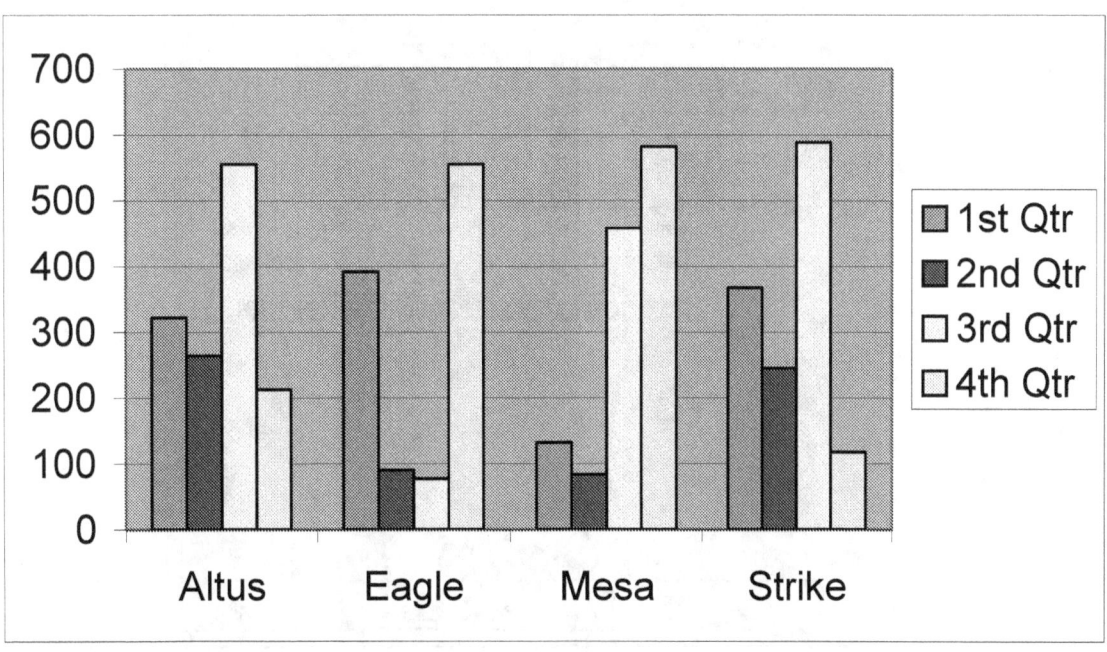 it will prompt you to select whether you'd like the headings displayed by columns or rows. Using the example below: if you select rows then the departments will be displayed in the X AXIS and the quarters will be listed in the LEGEND. If you select columns then the opposite will occur. You may also give your chart a title.

	Altus	Eagle	Mesa	Strike
1st Qtr	322	392	133	367
2nd Qtr	265	91	84	245
3rd Qtr	555	78	458	588
4th Qtr	213	555	582	118

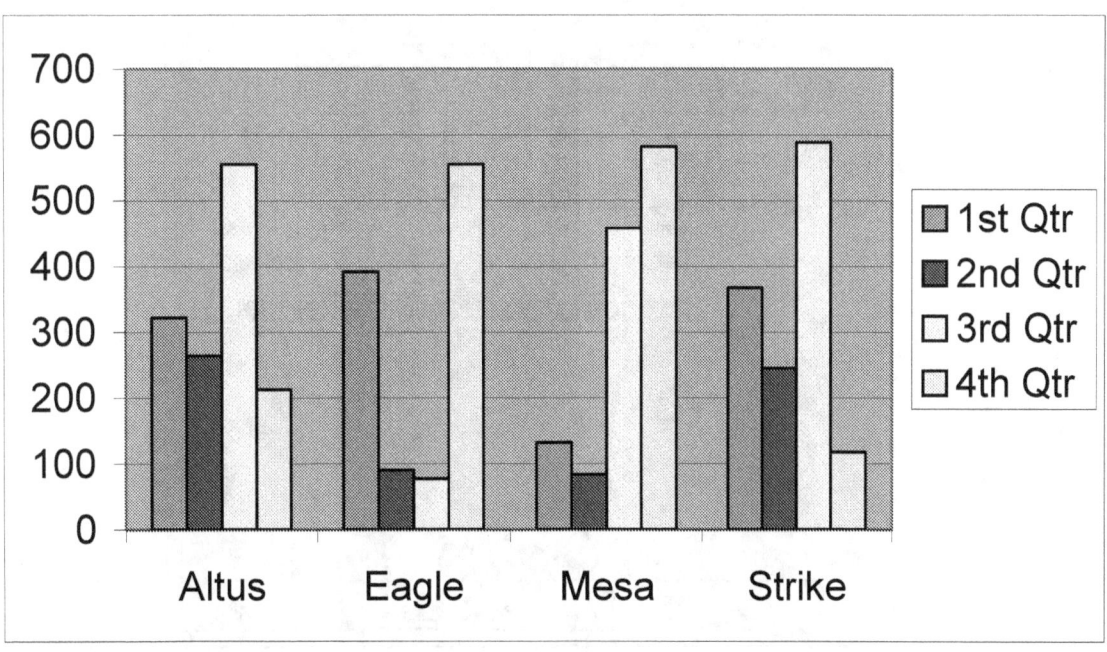

STEPS TO CREATING A CHART

STEP 1

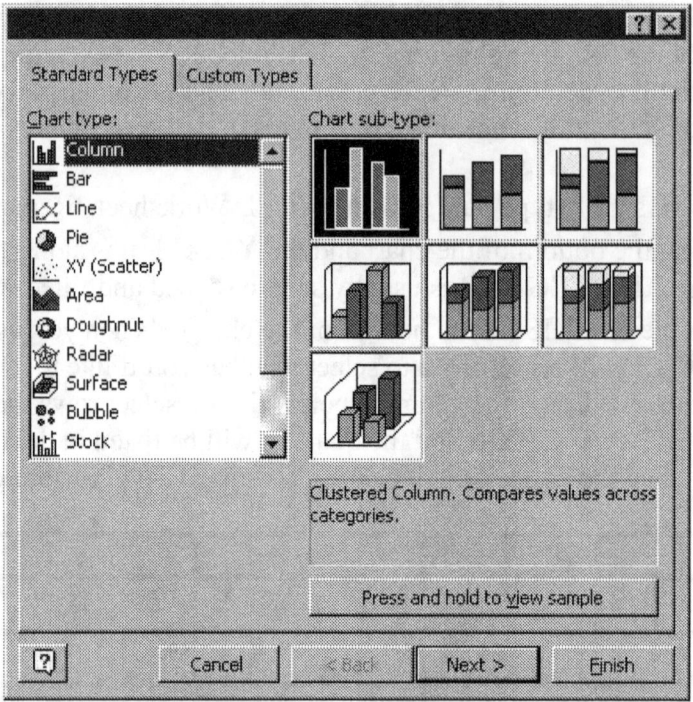

STEP 2

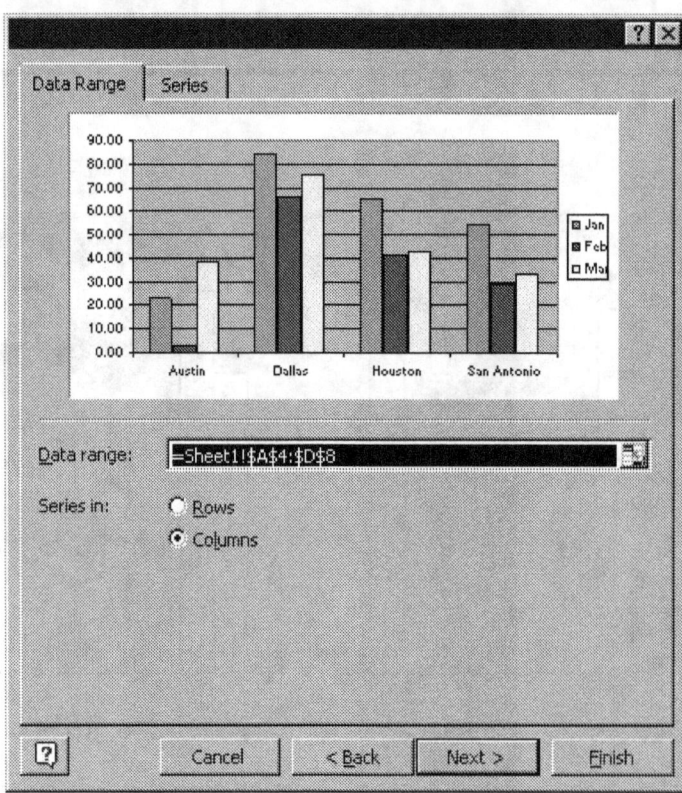

STEP 3

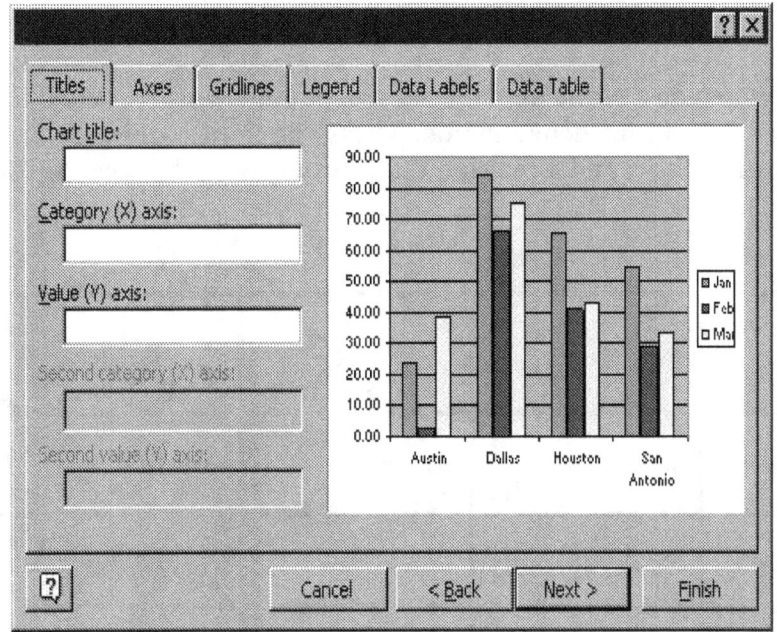

STEP 4

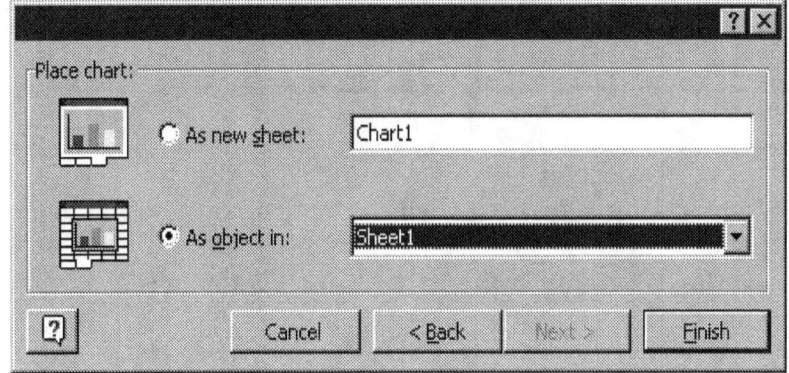

NOTES:_____

_____

_____

_____

_____

SALES CHARTS

In this Exercise you are the manager of a department store. Your task is to analyze the sales of four departments. Using the figures given below, you will create a column chart and a pie chart to compare sales between the four departments.

2003	Men's Casual	Women's Casual	Petite Casuals	Children's Casual
1st Qtr	$6,600	$6,000	$6,000	$10,855
2nd Qtr	$8,250	$21,345	$14,000	$13,575
3rd Qtr	$20,995	$9,000	$16,375	$11,945
4th Qtr	$12,750	$14,675	$10,550	$16,629

1. After looking at your sales chart, which department had the highest overall sales?

2. Notice anything troubling in the 1st, 2nd, and 3rd quarters?

NOTES:_____

_____

_____

_____

_____

Now use the data from the 3rd quarter to create a pie chart. See model below:

2003	Men's Casual	Women's Casual	Petite Casuals	Children's Casual
1st Qtr	$6,600	$6,000	$6,000	$10,855
2nd Qtr	$8,250	$21,345	$14,000	$13,575
3rd Qtr	$20,995	$9,000	$16,375	$11,945
4th Qtr	$12,750	$14,675	$10,550	$16,629

Pie Chart Model

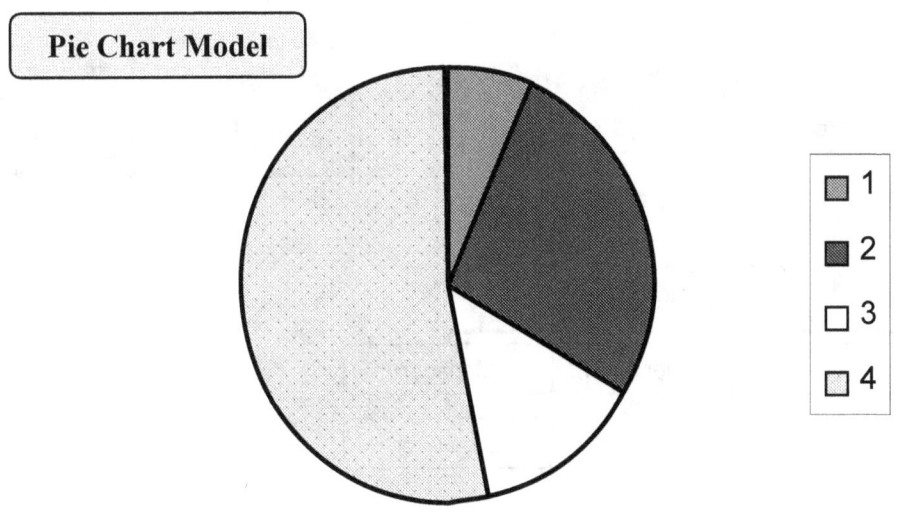

NOTES:_____

_____

_____

_____

_____

ORGANIZATIONAL CHARTS

Organizational charts are diagrams that depict the hierarchy of an organization or company. Organizational charts show the layout of the chain of command from top to bottom.

To create an organizational chart click on the INSERT command then select Diagram OR click the click the insert diagram icon. A new window will appear:

- Next double-click on the first selection in the diagram gallery.
- A generic diagram will appear
- Add names to your chart by clicking in the Area marked "Click to add text"
- On the Organization Chart Menu bar click the down arrow next to each option to adjust the overall layout or levels in the chain of command.

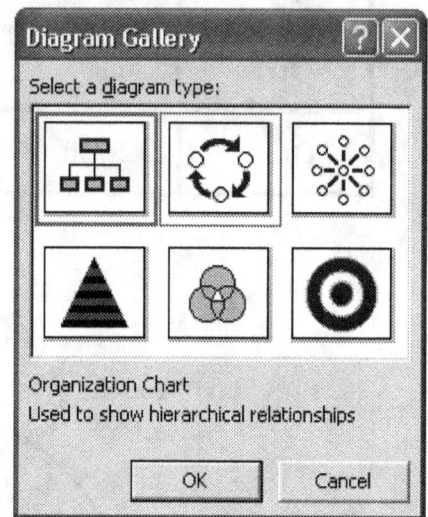

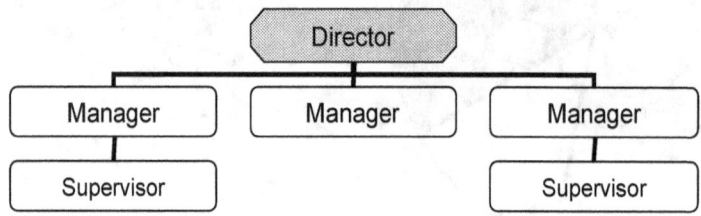

- To change the shape of the tabs within the chart, click on the frame of the in the chart.
- Click the Draw button located in the bottom right corner of the window.
- Select CHANGE AUTO-SHAPE
- A new menu will appear, select the shape that you prefer.
- You may also change the color of the shapes by clicking on the frame, then clicking the down arrow of the fill color icon and selecting your preferred color.

LINKING OBJECTS

Linking objects is a way of establishing a link to other programs or objects embedded in the same document.

LINKING DATA AND CHARTS
For this exercise you will need to have Excel and PowerPoint open.
Complete the following exercise: In Excel and type the following data.

	Altus	Eagle	Mesa	Strike
1st Qtr	322	392	133	367
2nd Qtr	265	91	84	245
3rd Qtr	555	78	458	588
4th Qtr	213	555	582	118

- Now select all of the data
- Then click the copy command
- Now switch windows to PowerPoint
- Select the icon to open a new slide (looks like a white piece of paper)
- Select the chart slide
- Double click to add chart
- When the Datasheet opens select appropriate number of cells for this exercise 5 columns and 5 rows.
- Select the EDIT command and Paste Link into the cell range
- Close the Data sheet

When you're done change the data in the Excel spreadsheet then view the chart on your PowerPoint slide, the chart should've changed as well.

HYPERLINKS

	Altus	Eagle	Mesa	Strike
1st Qtr	322	392	133	367
2nd Qtr	265	91	84	245
3rd Qtr	555	78	458	588
4th Qtr	213	555	582	118

To insert a HYPERLINK: Select the data to be linked; click on the INSERT HYPERLINK icon **Select PLACE IN THIS DOCUMENT, specify the sheet where the data will be linked to OR link data to a different document.**

CASH JOURNAL LINKING OBJECTS EXERCISE

Let's link two standard bookkeeping tools together; a monthly cash journal and balance sheet. Use separate worksheets. You may want to use the FREEZE PANES option to continue viewing the headings. After the template has been created you can copy it to a new worksheet, change the header, and plug-in data for the following month.

		Cash Journal for August			
Date	Check or Invoice #	Item Details	Debit (Expense)	Credit (Income/ Sales)	Net Accounts Receivable
8/2	001	Postage Fed Extra	$42.00		
8/5	002	Brochures	$125.00		
8/7	003	ACU Repair	$259.00		
8/7	4100	Boston Company		$450.00	
8/8	4101	Chiles Co		$550.00	
8/9	4103	Kane Co		$475.00	
8/9	004	Office Supplies	$70.00		
8/9	005	New Thermostat for ACU	$600.00		
8/12	4104	Hollings Estates		$1500.00	
8/14	4105	Bridge Homes			$2575.00
8/14	4106	Calloway Homes		$2500.00	
8/16	4107	Hollings Estates (new carpet)			$7600.00
8/17	006	Building Supplies-Lowell	$2000.00		
8/20	007	New Computer	$1500.00		
8/21	4108	Casey Homes		$2500.00	
8/22	008	New Truck Tires	$420.00		
8/25	4109	Hollings Apartments		$5500.00	
8/26	4110	Sundown Apartments		$5500.00	
8/27	009	Building Supplies	$8500.00		
8/28	010	Utilities	$275.00		
8/31	011	Rent	$3500.00		
Total Expense & Income August (Sum Columns)			Expenses FOMULA	Income FORMULA	Acct Receivable FORMULA

Next you'll set up a balance sheet.

BALANCE SHEET

You will link the sum of August's Income and Accounts Receivable from the Cash Journal to the Balance Sheet. When linked figures change in the Journal, they will also change the linked items in the Balance Sheet. Once this basic system is setup, it's simply a matter of accurately inputting the data and general maintenance. Ready? Let's get started.

Balance Sheet for August (ending Aug 31)			
Assets (company owned)		**Liabilities** (debts)	
Cash in the Bank	**$8,052.45**	September's Rent (unpaid)	**$3500.00**
Income (Sales) From August	Go the worksheet that has the Cash Journal; select the summed income from Cash Journal. **DO NOT DOUBLE CLICK if your cursor is blinking in the cell, try again.** Click on the EDIT command select COPY; at the bottom of screen go back to the Balance Sheet; click on the EDIT command; select PASTE SPECIAL; on the bottom left corner of that menu **SELECT PASTE LINK.**	Unpaid balance for heavy equipment	**$9000.00**
		Estimated Taxes	**$1200.00**
		Unpaid balance for portable lighting	**$500.00**
Heavy tools and equipment	**$20,000.00**		
Office Furniture	**$2000.00**		
Net Accounts Receivable	**PASTE LINK** Sum from Cash Journal	**Total Liabilities:**	$FORMULA SUM Liabilities
Total Assets	$FORMULA SUM Assets	**Net Income**	FORMULA Total Assets minus Total Liabilities

NOTES:_____

_____

COMPARATIVE MONTHLY BALANCE SHEETS

Now you will <u>Paste Link</u> the sum of each month's respective data from each month's Balance Sheet to the associated location within the Comparative Balance Sheet. Only Past Link data under the month of August. Only SUM items in shaded areas. Once this system is setup, it's simply a matter of accurately inputting the data and maintenance. Ready? Let's continue.

J. J.' s Construction Comparative Monthly Balance Sheets 2003						
	August	September	October	November	December	YEAR END TOTAL
ASSETS						
Cash in the Bank	PASTELINK	$10,000	$15,000	$50,000	$60,000	$60,000
Income from Sales	PASTELINK	$30,000	$35,000	$10,000	$5,000	SUM =sum(B4+ F4)
Heavy tools and equipment	PASTELINK	$21,000	$28,000	$28,000	$30,000	$30,000
Office Furniture	PASTELINK	$2,000	$2,800	$2,800	$3,500	$3,500
Net Accounts Receivable	PASTELINK	$5,000	$1,000	$8,000	$8,000 balance adjusts with payment	$8,000
TOTAL ASSETS	SUM	$68,000	$81,800	$98,800	$106,500	SUM =sum(Aug: Dec)
LIABILITIES						TOTAL
Rent	PASTELINK	$3,500	$3,500	$3,500	$3,500	$3,500
Unpaid balance for heavy equipment	PASTELINK	$8,000	$6,000	$15,500	$14,000	$14,000 running balance
Estimated Taxes	PASTELINK	$1,200	$1,300	$1,500	$1,800	SUM
Unpaid balance for portable lighting	PASTELINK	$450	$300	$0	$0	$0 running balance
TOTAL LIABILITIES	PASTELINK	$13,150	$11,100	$20,500	$19,300	SUM =sum(Aug: Dec)
NET INCOME: (Total Assets minus Total Liabilities)						

YEAR END BALANCE SHEET

OK, look how far you've gotten. GREAT JOB! Give yourself a pat on the back. Now wiggle your fingers and shake out your hands, there's a little more work ahead.

In the next exercise you will use the Paste Link method to link data from the Comparative Monthly Balance Sheet. The totals from each category will be linked to the associated cell on the Year End Balance Sheet. Please don't forget to use cell references (B10) etc. Ready? Let's move on.

Balance Sheet for Year Ended December 31, 2003

Assets (company owned)		Liabilities (debts)	
Cash in the Bank	**$60,000**	January's Rent (unpaid)	**$ 3,500**
Income Year 2XXX	**$ PASTE LINK.**	Unpaid balance for heavy equipment	**$14,000**
		Estimated Taxes	**$ PASTELINK**
		Unpaid balance for portable lighting	**$ 0**
Heavy tools and equipment	**$ 30,000**		
Office Furniture	**$ 3,500**		
Net Accounts Receivable	**$ 8,000**	**Total Liabilities:**	**$ PASTELINK**
		Net Income:	Assets minus
Total Assets	**$ SUM**	**FORMULA**	Liabilities

COMPARITIVE YEAR END BALANCE SHEETS

You will <u>Paste Link</u> the TOTAL from each year's respective cell on the Comparative Monthly Balance Sheet to the associated cell within the Comparative Year End Balance Sheet. You will only Paste Link data for the first year; the following year's information have been provided. Note that this model does not include a column titled TOTAL. OK. Let's press on.

J. J.' s Construction Comparative Year End Balance Sheets					
	2003	2004	2005	2006	2007
ASSETS					
Cash in the Bank	PASTELINK	$80,000	$95,000	$50,000	$60,000
Income From Sales	PASTELINK	$40,000	$35,000	$10,000	$5,000
Heavy tools and equipment	PASTELINK	$30,000	$50,000	$100,000	$230,000
Office Furniture	PASTELINK	$3,500	$4,800	$5,800	$5,800
Net Accounts Receivable	PASTELINK	$5,000	$1,000	$8,000	$8,000
TOTAL ASSETS	PASTELINK	$226,500	$185,800	$173,800	$308,800
LIABILITIES					
Unpaid Rent	PASTELINK	$3,500	$3,500	$3,500	$3,500
Unpaid balance for heavy equipment	PASTELINK	$28,000	$46,000	$58,500	$140,000
Estimated Taxes	PASTELINK	$14,200	$22, 300	$25,500	$50,800
Unpaid balance for portable lighting	PASTELINK	$5,000	$0	$0	$0
TOTAL LIABILITIES	PASTELINK	$13,150	$11,100	$20,500	$19,300
NET INCOME (assets-liabilities)	FORMULA Example =SUM(B10-B17)	FORMULA Example =SUM(C10-C17)	Repeat in to the right →		

INCOME STATEMENT

Time to cool down with some simple equations. Let's take a moment to reflect on the basics. Change the font color to whatever pleases you. Bold the titles under the items column. Italicize the headings of each column. Right align all numerical figures.

J. J's CAFÉ Income Statement Year Ended December 31, 2XXX		
ITEM	*EXPENSES*	
Gross Sales		$500,000
Cost of Goods Sold	$365,000	
Gross Profits		FORMULA =sum(Sales-goods)
Administrative Expense	$90,000	
Depreciation Expense	$40,000	
Operating Income		FORMULA =sum(gross minus admin minus depreciation)
Interest Expense	$45,000	
Earnings Before Taxes		FORMULA =sum(Operating Income minus Interest Expense)
Taxes	$30,000	
Earnings After Taxes		FORMULA =sum(Earnings Before Taxes minus Taxes)

Perhaps J.J. should stick to the construction business.

COMPARATIVE INCOME STATEMENT

Now you'll take a look at some of the café's quarterly data. Average the quarterly sales, cost of goods, and administrative expenses of J. J's Café.

J. J's CAFE Income Statement Comparative Quarterly Income Statement 2003				
	1st Qtr	2nd Qtr	3rd Qtr	AVERAGE
Gross Sales	$253,880	$149,650	$96,470	FORMULA
Cost of Goods Sold	$125,055	$121,550	$118,395	FORMULA
Gross Profits				
Administrative Expense	$40,000	$30,000	$20,000	FORMULA
Depreciation Expense				
Operating Income				
Interest Expense				
Earnings Before Taxes				
Taxes				
Earnings After Taxes				

J. J's Café has a real problem. It looks like the café began to lose money in the second quarter. However, the café's cost of goods remained high, in spite of the drop in sales.

WEB PAGE DESIGN

Many business owners have elected to advertise their products on a website. Having someone else create a web page or the entire website for your business can be costly. If you subscribe to the internet, most providers will give you a webpage for free. So to save money, do-it-yourself. In the next section you will create a simple layout for a webpage. Ready? Let's rock.

*Be mindful of where your page breaks are. Remember that they look like ---- perforated lines at the bottom and right side of your worksheet. On an unformatted worksheet page breaks will be positioned at the bottom of row 52 and the right side of the "I" column. The page breaks will move as you adjust the size of your columns and rows. **Data that is entered outside of the page breaks will appear on a separate page.** To set-up your webpage you'll need:

1. Internet service (preferably one that offers a free webpage), but .com addresses can be purchased.
2. The name of your business_____
3. Products and information that you want displayed_____

Let's create a webpage for J. J's Café. That place could use one. Open a new workbook. INSERT a new worksheet (you should now have a total of 4 worksheets). Remember to move the 4th worksheet to the fourth position (discussed on page 18).

1. In cell D1 type: J. J's Café (20 pt font size, Arial Black as the font) HIT ENTER
2. Click on cell D1 again and merge and center text……...**Merge-and-center icon**
3. In cell D2, type Voted "Best Burgers in Town!" (16 pt, Times New font) ENTER
4. Click on cell D2 again, **bold** and merge-and-center the text.
5. In cell A4 type: OUR MENUS (12 pt, **bold**, Arial font)
6. In cell C4 type BREAKFAST, D4 LUNCH, and E4 type DINNER (12 pt, **bold**, Arial)
7. Center the text in each cell. Don't forget to widen your columns.
8. Click on cell **C4**, next click the insert hyperlink icon
9. Select "Place in this document" select **sheet 2**, click OK
10. **D4** will be hyperlinked to **sheet 3** and **E4** will be hyperlinked to **sheet 4**
11. **To hyperlink to a specific cell: Select the cell to be linked; click the Insert Hyperlink icon; select Place In This Document; select the worksheet that you would like to link to; TYPE THE SPECIFIC CELL REFERENCE (B14 for example); click OK**
12. Go back to sheet 1, **INSERT CLIPART** (prefer motion clip) picture of a cook
13. Click VIEW command, select page break preview (page now breaks at column "F")
14. Return to Normal View
15. Cell C16 type: **OPEN 7 DAYS 6 to 6** (merge-and-center, **bold**, 12 pt, Arial)
16. Cell C17 type: **We are located at:** (merge-and-center, **bold**, 12 pt, Arial)
17. Cell C18 type: **23 Tasty Lane** (merge-and-center, **bold**, 12 pt, Arial)
18. Cell C198 type: **Snackville, USA 00819** (merge-and-center, **bold**, 12 pt, Arial)
19. Cell C20 type: **808-555-1616** (merge-and-center, **bold**, 12 pt, Arial)

WEB PAGE DESIGN CONTINUED

20. Use your mouse to highlight columns A through F, and rows 1 through 23
21. To highlight specific cells, click on cell A1, hold the left button of the mouse down, WITHOUT LETTING GO OF THE BUTTON, sweep over to the "F" column and down to row 23, release the button.
22. On the drawing toolbar, (bottom of window) click the down arrow next to the paint can.
23. Select the lightest shade of blue.
24. Click Worksheet 2, In cell D1 type: **Breakfast** (merge-and-center, **bold**, 20 pt, Arial)
25. In cells A4, A5, A6 type: Denver Omelet, Pancakes and Sausage, 2 Eggs Any Style
26. Click Worksheet 3, In cell D1 type: **Lunch** (merge-and-center, **bold**, 20 pt, Arial)
27. In cells A4, A5, A6 type: Big Burger, Turkey Burger, Sour Dough Burger
28. Click Worksheet 4, In cell D1 type: **Dinner** (merge-and-center, **bold**, 20 pt, Arial)
29. In cells A4, A5, A6 type: Fried Turkey Platter, Lemon Crumb Cod, Meat Loaf Plate
30. **Click the FILE command, select SAVE AS WEB PAGE**
31. Save in **My Documents** or **Personal**
32. **Select ENTIRE WORKBOOK**
33. **DO NOT ADD INTERACTIVITY** (Interactivity allows others to change the formatting)
34. **Name the File CAFÉ WEB PAGE; Click save**
35. **Click the FILE command, select WEBPAGE PREVIEW, Check your links**
36. Upload and publish your webpage based on the guidelines of your Internet Service Provider

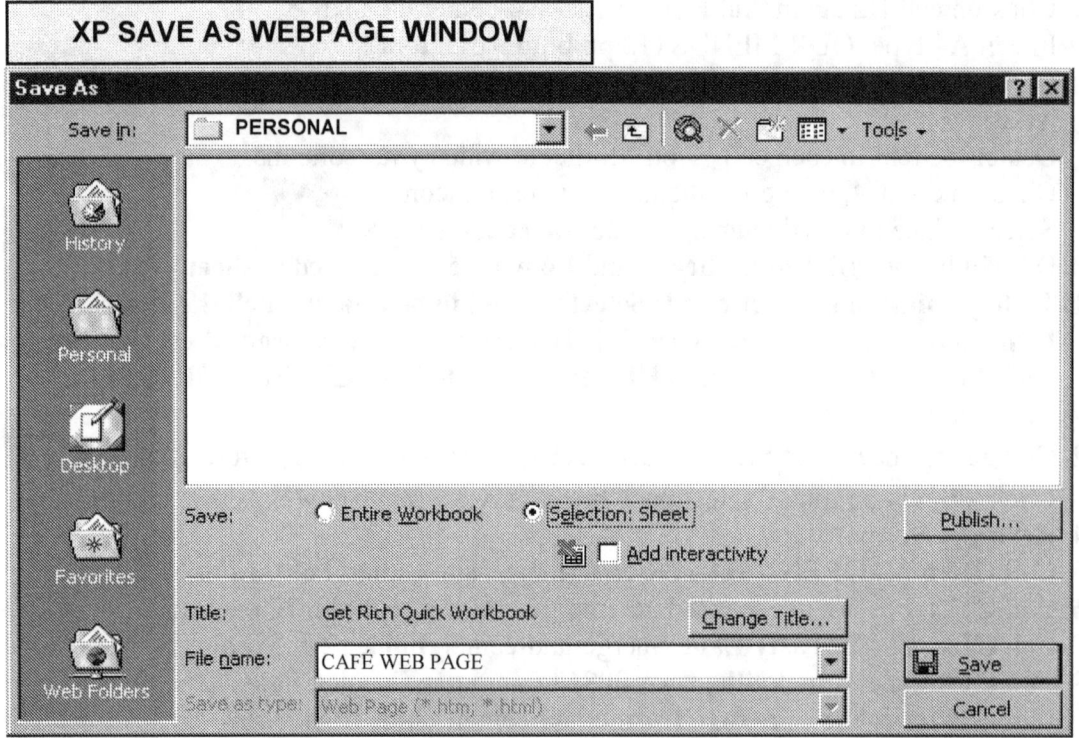

CREATING FORMS

Just for fun, let's touch on how to create forms. Creating forms in Excel can be challenging, but not impossible. Follow the steps below and decide if this is a useful tool for you.

TO CREATE A FORM OR QUESTIONNAIRE OR TEST

1. Open a new file
2. Merge and center the title of the form
3. Select VIEW from command bar, select toolbars, then FORMS and CONTROL TOOLBOX
4. Type in questions
5. **To insert an OPTION BUTTON: Click the icon labeled GROUP BOX ( ), draw a box, then click the icon labeled OPTION BUTTON, draw to desired size, edit text by selecting the DESIGN MODE icon ( ) right click on text area.**
6. Insert a comment to further explain specific words or points. (**Right click to edit).**
7. Protect the workbook with a password by selecting TOOLS, then PROTECTION, and then choose the item you'd like to protect, and enter a password.
8. **White worksheet: Clear gridlines headers by selecting TOOLS, then options, and uncheck the boxes marked gridlines and row and column headers** (you can also eliminate column/row headers or add gridline color).
9. To edit the size and with of cells without gridlines select the cell, click on FORMAT on main menu command bar, then COLUMN or ROW, type in desired width or height.

Create a Questionnaire titled: Employment Questionnaire

Check the preferred box

Full-time____   Part-time____   First Shift____   Second Shift_____

NOTES:_____

_____

_____

_____

_____

ANSWER KEY

EXERCISE #3

ADDITION
4
30
255
132
2066

SUBTRACTION
1
139
2445
553
-112
112

MULTIPLICATION
870
174
132
40920
137500
7770

DIVISION
4
40
10.75
19.74
33
409

AVERAGE
85
70.66667
89
96
71.66667
87.33333

EXERCISE #4

Manager	$1600.00
Supervisor	$960.00
Admin	$480.00
Admin	$490.00
Janitor	$720.00
Trainee	$460. 00
Trainer	$1200.00
Trainer	$980.00
Trainer	$1120.00
Trainer	$560.00

PURCHASES
Ending Balance $17380.00

SALES
Ending Balance $119880.00

BUDGET
Total costs $106570
Estimated capital $73,430

CASH JOURNAL
Debit $17,291.00
Credit $18,975.00
Accounts Rec. $10,175.00

BALANCE SHEET
Total Assets $59,202.45
Total Liabilities $14,200.00
Net Income $45,002.45

GLOSSARY

A

Ascending Order Data arranged in low to high or A to Z order.

Average Summary of data commonly called the mean.

B

Bar Chart A graphic depiction of data presented in a series of horizontal bars.

C

Cell A cell identifies an individual reference in the workbook (A1 or F17 for example)..

Charts Graphic depictions of data and figures illustrated in columns, bars, lines, pie, scatter, etc.

D

Data Information within a cell.

F

Filter Function that locates and isolates certain data within a worksheet.

Formula A mathematical equation that accomplishes calculations within a workbook.

Freezing Panes Permits continuous viewing of the headings at the top of each column while at the same time allowing you to input or view data listed beneath the headings.

H

Hyperlinks Allows you to access a page or file by clicking on specifically linked data.

I

Interactivity Will allow others to change the data and formatting in a webpage.

L

Linking Objects Linking objects allows you to establish a link to other related programs or objects embedded in the same document.

O

Operators Are used to express and Calculate equations (+, -, *, /, =).

R

Range Reference (:) Refers to or identifies a range of cells (A1:A10) (A1:G1).

S

Sort Alphabetizes or categorizes a list of data.

W

Workbook The file in which you work and store your data (usually a collection of worksheets.

Worksheet A sheet or page within a workbook.

Index

ACKNOWLEDGEMENTS

- To my dear sister Lisa T., thank you for always being there for me. Here's your hug.

- To my charming daughter Soummer, may your fondest dreams and prayers come true. Love you Kidd.

- To my talented nephew Damarko, the kid who believes he can reach beyond the stars, smart guys like you always do. Love you Sharko.

- To little 4 year old P.J. No you can not go to the grocery store and buy a new tooth for one dollar. God will give you a new tooth for free. You'll just have to be patient.

- To my brother Don, please know that you hold a very special place in my heart.

- To my Grandmother Otherine, we rise because of your foundation and those sweet potato pies filled with love. Love you Grandma.

- To my aunt Jackie, there are not enough words to describe your generous heart; Love you.

- To Boyd H., I am so proud that you are such a great father to our daughter. We were childhood sweethearts and you grew up to be one heck of a dad.

- To Laureen Primus my friend, my sister, I admire your spirit.

- To my hardest working uncle Cedric, I'm so happy to call you my uncle.

- To my first cousin Al., thank you for everything.

- To my uncle Nate, thanks for your sweetness.

- To my friend David "Dutch" Dudley you mean the world to me. Thank you Dutch.

- To Lisa Campbell, my good friend and colleague, thanks for being such a wonderful friend.

- To my friends Mary and Tony, thanks for all of your support. Man I love you guys!

- To my good friends Wil Young and Jeff Jumper, you guys really make me smile.

- To my uncle Haywood

- To my aunt Ronnette may you be blessed with a voice to tell your story.

- To my friend Nathan AKA "Boots" if I haven't told you, "thank you."

- To Connie and Larry G., thanks for being there.

- To my sixth grade teacher Mr. N. Hudson of Herman Elementary School, you really made a difference. I wish that I could clone you a million times so that we could people the world with wonderful teachers.

- Special thanks to Yvette R., my prayers are with you.

References

Worm, © 1999, Corel. All rights reserved.

Icons and Windows, © 1999, 2000, Microsoft. All rights reserved.

Minding Your Business With EXCEL

Don't let the size of this book fool you. It's packed with tips, exercises and illustrations. In Basics to Business Minding Your Business with Excel, author S, D. Moore illustrates simple applications for using Excel in a small business.

The author reveals a step-by-step guide to learning the basic and business applications of Excel. Basics to Business lays out a simple, straightforward path without being too elementary or complex. This book is a must have for the savvy entrepreneur.